What is Scie

by Marcia S. Freeman

ROURKE CLASSROOM RESOURCES
★ *The path to student success*

Science is looking closely at things.

2

Science is comparing things.

Science is collecting things.

Science is counting things.

Science is measuring things.

Science is wondering about things.

Science is trying things out.